FRUITS YOU LOVE TO EAT

BANANAS

AMY CULLIFORD

A Crabtree Roots Book

Crabtree Publishing
crabtreebooks.com

School-to-Home Support for Caregivers and Teachers

This book helps children grow by letting them practice reading. Here are a few guiding questions to help the reader with building his or her comprehension skills. Possible answers appear here in red.

Before Reading:

- What do I think this book is about?
  - *I think this book is about how to grow bananas.*
  - *I think this book will tell me it's healthy to eat a banana every day.*

- What do I want to learn about this topic?
  - *I want to learn why some bananas are large and others are small.*
  - *I want to learn why some bananas have little brown spots.*

During Reading:

- I wonder why...
  - *I wonder why bananas come in so many colors.*
  - *I wonder why some bananas taste sweet and others do not.*

- What have I learned so far?
  - *I have learned that bananas are fruits.*
  - *I have learned that bananas grow from plants.*

After Reading:

- What details did I learn about this topic?
  - *I have learned that all bananas have peels.*
  - *I have learned that bananas come from big purple buds on banana plants.*

- Read the book again and look for the vocabulary words.
  - *I see the word* ***buds*** *on page 6 and the word* ***peels*** *on page 10. The other vocabulary words are found on page 14.*

Bananas are **fruits**.

Bananas come from **plants**.

The plants grow big **purple buds** that turn into bananas.

Bananas can be green, brown, purple, or yellow.

All bananas
have **peels**.

I like to eat bananas! Yum!

Word List

Sight Words

all	come	I	to
are	eat	into	turn
be	from	like	yellow
big	green	or	
brown	grow	that	
can	have	the	

Words to Know

bananas

buds

fruits

peels

plants

purple

35 Words

Bananas are **fruits**.

Bananas come from **plants**.

The plants grow big **purple buds** that turn into bananas.

Bananas can be green, brown, purple, or yellow.

All bananas have **peels**.

I like to eat bananas! Yum!

FRUITS YOU LOVE TO EAT

BANANAS

Written by: Amy Culliford
Designed by: Rhea Wallace
Series Development: James Earley
Proofreader: Melissa Boyce
Educational Consultant: Marie Lemke M.Ed.

Photographs:
Shutterstock: Nataly Studio: cover; Maks Narodenko: p. 1; Tatjana Baibakova: p. 3; Wang Sling: p. 5; thanannopthanoch thongnam: p. 7; Photoography: p. 9; BigcStudio: p. 11; naluwan: p. 13

Crabtree Publishing

crabtreebooks.com 800-387-7650

In Canada: We acknowledge the financial support of the Government of Canada through the Canada Book Fund for our publishing activities.

Printed in the U.S.A./072023/CG20230214

Published in Canada
Crabtree Publishing
616 Welland Ave.
St. Catharines, Ontario
L2M 5V6

Published in the United States
Crabtree Publishing
347 Fifth Ave
Suite 1402-145
New York, NY 10016

Library and Archives Canada Cataloguing in Publication
Available at Library and Archives Canada

Library of Congress Cataloging-in-Publication Data
Available at the Library of Congress

Hardcover: 978-1-0398-0974-1
Paperback: 978-1-0398-1027-3
Ebook (pdf): 978-1-0398-1133-1
Epub: 978-1-0398-1080-8